TIME SAVING Teaching with AI

THE BURNED OUT TEACHERS' GUIDE TO:

CRAFTING ENGAGING LESSONS WITH CHAT GPT

Includes Prompts and Sample Lesson Plans

BY CELESTE PAYNE

www.celesteempowers.com

Special discounts are available on quantity purchases by corporations, associations, and others. For details, contact 855.213.4345 or at www.celesteempowers.com.

ISBN: 978-1-930276-01-7

Printed in the United States of America

First Edition, 2024

TABLE OF CONTENTS

ABOUT ME

I'm Celeste Payne, an educator, author, and artist that uses arts and technology to help people from all walks of life de-stress, save time, and discover innovative solutions to barriers that may be hindering their success. I am a certified art teacher, and award-winning graphic designer that empowers small businesses in graphic design education and supports my clients through online courses and coaching. I'm a multiple Amazon best-selling author and an impactful presenter at empowerment workshops and conferences. Through in-person and virtual courses in AI, CANVA, and the arts, I have been able to inspire and empower people from across the globe.

I am founder of Arts Empowered, Inc., a Florida non-profit dedicated to enriching lives and fostering community engagement through the arts, addressing the need for accessible art education, art as self care, and arts technology.

I am also founder of CelesteEmpowers, a professional and personal empowerment and coaching firm.

CelesteEMPOWERS
FANNING THE FLAMES OF CREATIVITY

www.celesteempowers.com

YOUR CREATIVE EDUCATION CONSULTANT
CANVA, AI
GRAPHIC DESIGN
• Online Courses
• Workshops/Webinars
• Technical Support
• Consulting
COMMUNITY
Canva '24
CANVASSADORS
ARTS
EMPOWERED
Celeste EMPOWERS
FANNING THE FLAMES OF CREATIVITY
Celeste Payne
Educator, Graphic Artist, Author
celesteempowers.com

PREFACE

What is a ChatGPT Prompt?

A ChatGPT prompt is a specific instruction or query that you provide to ChatGPT, an AI language model developed by OpenAI. The prompt guides the AI in generating a response that aligns with your request. Essentially, it acts as a starting point for the AI to understand what information you are seeking or what task you need assistance with.

Key Elements of a ChatGPT Prompt

1. **Clarity and Specificity:** The prompt should clearly and specifically articulate what you want the AI to do. The more precise the prompt, the more accurate and relevant the AI's response will be.
2. **Context:** Providing context helps the AI understand the background or the specifics of a situation, making its responses more tailored and effective.
3. **Purpose:** Every prompt should have a clear purpose or goal, whether it's generating educational content, providing explanations, or creating assessments.

Uses of ChatGPT Prompts in Education

Teachers can use ChatGPT prompts in a variety of ways to enhance teaching and learning:

1. **Content Creation:** Teachers can use prompts to generate instructional materials, such as lesson plans, worksheets, or quizzes. For example, a prompt could be, "Generate a quiz on photosynthesis for a 9th-grade biology class."

2. **Student Engagement:** Prompts can be designed to create engaging activities or discussion questions that stimulate student thought and participation. An example might be, "Create five discussion questions that encourage critical thinking about Shakespeare's 'Macbeth.'"

3. **Personalized Learning:** Teachers can input prompts to develop customized learning activities that cater to students' individual needs and learning styles. For instance, "Suggest three differentiated strategies for teaching fractions to students with varying math abilities."

4. **Assessment:** Prompts can help in designing formative or summative assessments that are aligned with

educational standards and learning objectives. An example prompt might be, "Design a summative assessment for a unit on the American Civil War that includes both multiple-choice and essay questions."

5. **Feedback and Reflection:** Teachers can use prompts to generate tools for student feedback, self-assessment, and reflection activities that help students process their learning experiences. For example, "Create a reflection activity for students after completing a group project on environmental science."

How to Craft Effective ChatGPT Prompts

To craft effective ChatGPT prompts, teachers should:

- Be specific about what they want to achieve with the response.
- Include relevant details that might influence the AI's output, such as grade level, subject area, or specific learning goals.
- Phrase the prompt clearly, avoiding ambiguous language to minimize the risk of receiving irrelevant or incorrect information.

By understanding and utilizing ChatGPT prompts effectively, teachers can leverage this powerful AI tool to support and enhance their instructional practices, making their teaching more efficient and responsive to students' needs.

How do I access ChatGPT

To access ChatGPT, visit OpenAI's official website at www.chatgpt.com. OpenAI offers several subscription levels, including a free tier which provides basic access with limited usage, and paid subscription plans for more extensive features and higher usage limits. These plans are suitable for users requiring more robust access or additional capabilities, including educators who may benefit from educational discounts or special access to advanced features.

OpenAI sometimes offers plans with enhanced tools and resources specifically designed for educational use, helping teachers and institutions effectively integrate AI into their curricula. For the latest information on such offerings, it's best to check the OpenAI website directly or contact their support.

ChatGPT is versatile and can be accessed using various devices, including your phone, tablet, laptop, or desktop computer. This flexibility ensures that you can use the service at home, in the classroom, or on the go, adapting to different teaching and learning environments. After selecting the appropriate subscription level and logging in, you can start using ChatGPT by typing your queries or prompts into the chat interface provided on the platform. This setup allows educators and other users to fully utilize advanced AI features in a way that suits their needs and enhances their educational or professional endeavors.

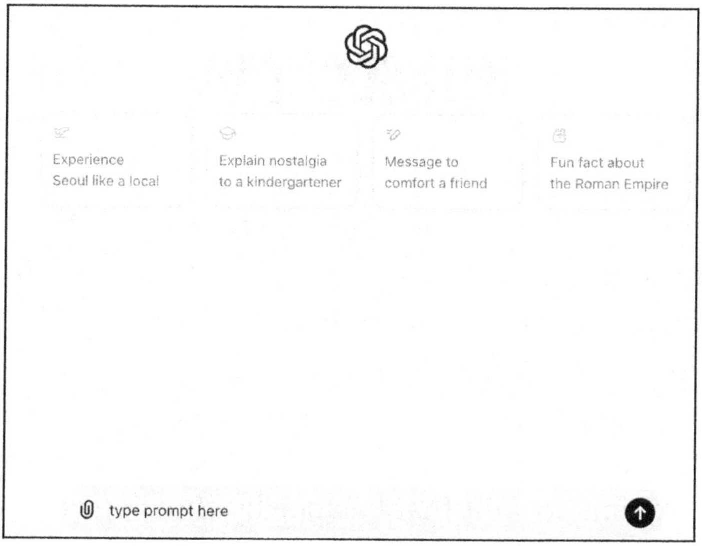

INTRODUCTION

Welcome to "Time Saving Teaching: With AI: The Burned Out Teachers' Guide to Creating Engaging Lessons with ChatGPT" ! As an educator with 12 years of experience, I've seen firsthand how teaching can be both incredibly rewarding and immensely demanding. One of the biggest challenges we face is the time-consuming process of writing detailed lesson plans, which often takes our focus away from the actual teaching. This ebook aims to shift that balance by showing you how to leverage AI technology to streamline your lesson planning process, allowing you to focus more on teaching and less on paperwork.

The introduction of AI technology, particularly Chat GPT, has revolutionized my approach to lesson planning. This powerful tool has allowed me to streamline the planning process, enabling me to spend more time on what truly matters—engaging with my students and delivering high-quality instruction. Chat GPT is an advanced AI technology that can assist in generating lesson content, offering suggestions, and providing a structured framework for your teaching plans. Its role in lesson planning is to act as a supportive tool that enhances your creativity and efficiency, making the planning process smoother and more intuitive.

The purpose of this ebook is to guide you through the practical applications of Chat GPT in creating comprehensive and effective lesson plans. We'll explore how this AI tool can help you align your lessons with educational standards, use your school's templates, build from existing scope and sequences, and incorporate key teaching strategies such as Bell Ringers and the I Do, We Do, You Do model.

Together, we'll explore how to use your school's templates, build lessons from existing scope and sequence, and tailor your plans to meet the diverse needs of your students through accommodations and differentiation. By the end of this book, you'll be equipped with the knowledge and tools to enhance your teaching practice, making lesson planning a breeze.

The benefits of using AI in education are profound. With Chat GPT, you can save valuable time, reduce the stress of lesson planning, and ensure that your lessons are engaging and aligned with your educational goals. By embracing this technology, you'll be able to dedicate more energy to connecting with your students and fostering a dynamic learning environment.

Let's embark on this journey together and discover how AI can transform your teaching experience, making lesson planning not just manageable but enjoyable!

CHAPTER 1: UNDERSTANDING EDUCATIONAL STANDARDS

What are Educational Standards?

Educational standards are specific, detailed descriptions of what students are expected to learn and be able to do at certain points in their education. These standards are developed by educational boards and organizations to ensure a consistent, high-quality education for all students across different grades and subjects. Standards are not curriculum, but rather the benchmarks for assessing student skills and knowledge that inform curriculum development.

Importance of Aligning Lessons with Standards

Aligning lessons with educational standards is crucial for several reasons. First, it ensures that teaching practices are focused on clear, measurable objectives that are recognized as important within the educational community. This alignment helps educators systematically drive students towards achieving key learning outcomes. Furthermore, standards-based education aids in maintaining high academic consistency across different schools and districts, ensuring all students have access to a quality education that prepares them for future educational and career opportunities.

How Chat GPT Can Help Incorporate Standards

Chat GPT can be a powerful tool in incorporating educational standards into lesson planning. Here's how it can assist:

1. By uploading or copying and pasting your desired standard(s) into your lesson planning prompts, you can have ChatGPT align your proposed lesson to these standards.

References from Educational Journals

For further reading and a deeper understanding of educational standards, the following references from respected educational journals are invaluable:

- Journal of Education Policy: Provides analyses of policy changes and their implications for educational standards.
- American Educational Research Journal: Offers insights into the research underpinning educational standards and their practical applications in the classroom.

- Teaching and Teacher Education: Focuses on how educators can effectively implement standards in their teaching practices.

Incorporating Chat GPT into your lesson planning not only helps align with educational standards but also enhances the effectiveness and efficiency of your teaching. As we continue to explore this tool's capabilities, it becomes clear that AI can play a significant role in elevating educational outcomes.

CHAPTER 2: UPLOADING AND USING SCHOOL TEMPLATES

Introduction to School Templates

As a teacher, one of the most critical tools at our disposal for creating structured and consistent lesson plans are templates. These templates are not just a set of guidelines but a framework that ensures each lesson meets the school's educational standards and teaching objectives. They typically include sections for objectives, materials needed, detailed activities, assessment methods, and reflection components. Utilizing these templates can greatly improve the efficiency of our lesson planning and ensure uniformity in the delivery of educational content across different classes and grades. The components of these templates can be uploaded as reference material to be included in your Chat GPT generated lesson plan

Step-by-Step Guide to Uploading Templates into Chat GPT

Integrating school templates into Chat GPT can significantly streamline the lesson planning process. Here's how you can do it:

1. **Prepare Your Template:** Begin by ensuring your template is in a digital format, preferably a text-based file such as a DOCX or a PDF. This preparation might involve digitizing a paper template if it's not already in an electronic form.
2. **Input into Chat GPT:** You can input the template into Chat GPT by simply typing out the structure or by using file upload features if supported. Explain each section of the template briefly to Chat GPT to help it understand the components it will work with.
3. **Initial Setup:** Once uploaded, instruct Chat GPT to recognize and categorize different parts of the template. For example, you can set commands like "Identify learning objectives section," "List all activity prompts," or "Outline assessment strategies."
4. **Saving Template in Chat GPT**: The Chat GPT tool so that you can reuse it with different content in the future. This step saves time and maintains consistency across all your lesson plans.

Customizing Lessons Based on Template Components

Once your template is set up in Chat GPT, customizing lessons becomes a straightforward task. Here's how to make the most of this setup:

1. **Generating Content:** Ask Chat GPT to generate specific content for each part of the template. For instance, input the topic or learning objective, and let Chat GPT propose activities and methods for assessment that fit within the template's structure.
2. **Adjustments and Tweaks:** Review the generated lesson plan to ensure it aligns with your students' needs and make adjustments as necessary. You can modify the activities for different learning styles or add additional resources.
3. **Feedback Incorporation:** After teaching the lesson, you can input any outcomes or feedback into Chat GPT to refine future lesson plans. For example, if an activity was particularly effective or needed improvement, you can note these observations to enhance or adjust the template's activity suggestions.
4. **Continuous Improvement:** Continuously use the feedback from each lesson to improve the template itself. Chat GPT can assist in optimizing the template over time, ensuring that it remains relevant and effective based on actual classroom experiences.

CHAPTER 3: CREATING LESSONS FROM EXISTING SCOPE AND SEQUENCE

Defining Scope and Sequence

In the context of education, "scope and sequence" refers to a comprehensive blueprint that outlines what is taught and the order in which topics are covered throughout an academic program or course. The scope details the breadth of content—what students will learn, including the skills and knowledge they are expected to acquire. The sequence specifies the order and timeline for when each part of the content should be taught, ensuring a logical progression of learning activities that build upon each other.

How to Input Scope and Sequence into Chat GPT

Incorporating your scope and sequence into Chat GPT can transform how you approach lesson planning by making it more aligned with your long-term educational goals. Here's how to integrate this effectively:

1. **Digitize Your Scope and Sequence:** If your scope and sequence are not already in a digital format, the first step is to convert them into a text or spreadsheet file. This conversion makes it easier to handle and manipulate the data.

2. **Break It Down:** Divide your scope and sequence into manageable parts, such as by units or chapters. This breakdown will help Chat GPT handle the information more effectively, allowing for targeted lesson creation.

3. **Upload to Chat GPT:** You can input the scope and sequence by typing it directly into Chat GPT or by uploading the file, if the platform supports this feature. Provide context to Chat GPT by explaining what each part represents, and clarify how detailed the lesson plans need to be.

4. **Set Parameters:** Specify any particular focus areas, methodologies, or constraints you want to adhere to based on your curriculum requirements. This step ensures that the

lessons generated by Chat GPT align with both your educational standards and teaching style.

Generating Lessons That Align with Your Curriculum Plan

With your scope and sequence set up in Chat GPT, you can start generating lesson plans that are tailored to your curriculum. Here's how to maximize the effectiveness of this process:

1. Generate Ideas: Ask Chat GPT to suggest lesson ideas that cover specific topics within your scope and sequence. You can request broad overviews or detailed activities depending on the depth of content you need.

2. Customize Content: Customize the generated ideas to fit your teaching style and the needs of your students. This customization might involve adjusting the complexity of the tasks, incorporating different teaching aids, or integrating cross-curricular links.

3. Integrate Assessments: Ensure that your lessons include appropriate assessments to evaluate student understanding. Chat GPT can help design quizzes, tests, and other assessment tools that align with the objectives outlined in your scope and sequence.

4. Iterate Based on Feedback: After conducting lessons, gather feedback and assess student performances to refine future plans. Feed this information back into Chat GPT to adjust the upcoming lesson plans, ensuring they remain responsive to student needs and effective in meeting learning goals.

5. Continuous Curriculum Alignment: Regularly review the alignment of generated lessons with your overall curriculum goals. Make adjustments to the scope and sequence input in Chat GPT as needed, to reflect changes in curriculum standards or educational objectives.

CHAPTER 4: INCORPORATING BELL RINGERS (DO NOWS)

Importance of Bell Ringers in Lesson Plans

Bell Ringers, or "Do Nows," are short, introductory activities or prompts that students engage with at the very start of a lesson. These activities serve multiple critical educational functions: they transition students from their previous activities into a learning mindset, set the tone for the class, activate prior knowledge, and can be used to assess understanding from previous lessons. Incorporating Bell Ringers ensures that classes start promptly and effectively, maximizing instructional time and reinforcing routine and expectations among students.

Examples of Effective Bell Ringers

Effective Bell Ringers are typically aligned with the day's lesson, straightforward, achievable in a short timeframe, and engaging enough to capture students' interest from the moment they sit down. Here are a few examples:

1. **Question of the Day**: Pose a relevant question related to the lesson's theme or a concept from a previous class. For instance, in a history lesson about the American Revolution, you might ask, "Why do you think the Boston Tea Party was a significant event in American history?"
2. **Quick Quizzes**: Start with a mini-quiz related to the previous day's material to assess retention and understanding. This can be especially useful in subjects like math or science, where continual practice is crucial.

3. **Journal Entry**: Encourage students to write a brief response to a prompt that connects with the day's objectives. For a literature class, you might prompt with, "Write about a time when you faced a challenge similar to that of the protagonist in our current book."

4. **Visual Analysis**: Present a relevant image or video and ask students to analyze it based on current learning topics. For example, in an art class, show a piece of

artwork and ask, "What emotions do you think this artwork is trying to convey?"

Using Chat GPT to Generate Engaging Bell Ringers

Chat GPT can be an invaluable tool in designing dynamic and effective Bell Ringers. Here's how to utilize it:

1. **Idea Generation**: Provide Chat GPT with the topic or focus of your upcoming lesson and ask it to generate several Bell Ringer ideas. Chat GPT can suggest questions, small projects, or critical thinking prompts that are tailored to your specific lesson objectives.
2. **Customization**: Tailor the generated ideas to better fit your students' needs or the specific context of your lesson. You might adjust the difficulty level, cultural relevance, or even the format (e.g., changing a written prompt to a discussion-based activity).

3. **Diversity and Inclusion**: Use Chat GPT to incorporate diverse perspectives and inclusive prompts. This can involve asking for Bell Ringers that touch on various cultural backgrounds, learning styles, and abilities, ensuring that all students feel represented and engaged from the start of the lesson.

4. **Feedback Integration**: After using a Bell Ringer, quickly gauge its effectiveness and student engagement. Feed

this information back into Chat GPT to improve and refine future Bell Ringers, making them even more engaging and relevant.

5. **Regular Updates**: Continuously update the pool of Bell Ringers by inputting new themes, topics, or educational trends into Chat GPT. This will keep your starter activities fresh and exciting, which is key to maintaining student interest and engagement.

Sample Chat GPT Prompts for Bell Ringers

1. **Quick Review Questions**:
    - **Prompt**: "Generate a list of five quick review questions on the topic of cellular respiration for a high school biology class. Each question should take no more than two minutes to answer."

2. **Warm-Up Problems**:
    - **Prompt**: "Create three math warm-up problems for an 8th-grade class that cover yesterday's topic of graphing linear equations."

3. **Journal Prompt**:
    - **Prompt**: "Provide a thought-provoking journal prompt related to the theme of leadership,

suitable for a middle school social studies class."

4. **Quote Analysis**:
    - **Prompt**: "Select a quote from Martin Luther King Jr. and formulate a Bell Ringer activity where students analyze the meaning and how it applies to today's society."

5. **Image Analysis**:
    - **Prompt**: "Find an iconic photograph from the American Civil Rights Movement and create a Bell Ringer activity where students discuss the impact of imagery in historical movements."
6. **Mind Teasers**:
    - **Prompt**: "Create a mind teaser related to logical reasoning that can be used as a Bell Ringer for a high school critical thinking class."

7. **Prediction Challenge**:
    - **Prompt**: "Design a Bell Ringer that asks students to make predictions about a story they are reading in English class, focusing on character development and plot twists."

8. **Historical Fact of the Day**:
    - **Prompt**: "Generate a 'Historical Fact of the Day' related to our current unit on World War II, and

suggest a discussion question to accompany the fact."

9. **Grammar Exercise**:
    - **Prompt**: "Create a quick grammar exercise focusing on the use of their, there, and they're, suitable for a language arts class."

10. **Vocabulary Builder**:
    - **Prompt**: "List five new vocabulary words related to the current science unit on ecosystems. Include definitions and use each word in a sentence."

These prompts are designed to engage students right at the start of class, tapping into various skills and subject areas while seamlessly connecting with the day's lesson or reviewing previous material. Using ChatGPT to generate these activities can save time and add variety to the daily routine.

CHAPTER 5:
THE I DO, WE DO, YOU DO MODEL

Explanation of the I Do, We Do, You Do Instructional Model

The "I Do, We Do, You Do" model is a structured approach to teaching that breaks down learning into three distinct phases, each critical to student understanding and success. This model is especially effective for introducing new material and skills, providing a scaffolded learning experience that gradually shifts responsibility from the teacher to the student.

1. **I Do**: This is the teacher-led instruction phase where the teacher demonstrates the skill or concept, providing clear examples and often thinking aloud to model the thought process behind the task.
2. **We Do**: During this phase, the teacher and students work together to practice the new skill. This guided practice allows students to try the task with the safety

net of immediate feedback and support from the teacher.

3. **You Do**: This final phase involves independent practice where students apply the skill on their own, demonstrating mastery and the ability to perform the task without direct teacher guidance.

Creating Lessons Using This Model with Chat GPT

Utilizing Chat GPT to plan lessons based on the "I Do, We Do, You Do" model can enhance the effectiveness of your instruction by ensuring each phase is thoughtfully prepared and tailored to students' needs.

I Do: Teacher-Led Instruction

- **Generating Content**: Ask Chat GPT to provide a clear explanation of the concept or a step-by-step demonstration of the skill. This could include important vocabulary, key points to emphasize, and common misconceptions to address.
- **Modeling Examples**: Request that Chat GPT generate specific examples or problems that you can solve in front of the class, demonstrating the process and decision-making involved.

We Do: Guided Practice

- **Collaborative Activities**: Use Chat GPT to create activities or problems that you can work through as a class. Ensure these activities are directly related to the "I Do" phase to reinforce learning.

- **Interactive Questions**: Have Chat GPT suggest questions that students can answer in pairs or small groups, facilitating discussion and deeper understanding.

You Do: Independent Practice

- **Practice Assignments**: Direct Chat GPT to develop practice assignments that students can complete on their own. These should challenge the students but not be so difficult that they feel frustrated without support.
- **Assessment Tasks**: Generate assessment tasks that allow students to demonstrate their understanding and mastery of the subject matter independently.

Examples of Lessons Using This Model

Here are a couple of examples of how a lesson might be structured using the "I Do, We Do, You Do" model:

1. **Math Lesson on Fractions**:

- **I Do**: The teacher explains the concept of fractions, using a pie chart to demonstrate how a whole is divided into parts.
- **We Do**: Together, the class works through several problems on adding fractions with like denominators, with the teacher guiding the process and students offering input.
- **You Do**: Students complete a worksheet on their own, solving problems that involve adding and subtracting fractions, which they submit for review.

2. **Science Lesson on Photosynthesis**:
    - **I Do**: The teacher describes the process of photosynthesis, detailing the roles of sunlight, water, and carbon dioxide. Diagrams and a short video are used for demonstration.
    - **We Do**: The class conducts a guided experiment with plants under different light conditions.
    - **You Do**: Students design their own experiment to test variables affecting photosynthesis and prepare a brief report or presentation of their findings.

SAMPLE PROMPTS FOR I DO, WE DO, YOU DO

I Do: Teacher-Led Instruction

1. **Explain a Concept**:
    - **Prompt**: "Generate a clear and concise explanation of the water cycle for a fifth-grade science class, including key terms such as evaporation, condensation, and precipitation."
2. **Demonstration**:
    - **Prompt**: "Provide a step-by-step demonstration on how to solve a linear equation. Include important points to emphasize and common mistakes that students should avoid."
3. **Visual Aids**:
    - **Prompt**: "Create a list of visual aids that can be used to explain the concept of ecosystems to a middle school class, emphasizing the interactions between different organisms and their environment."

We Do: Guided Practice

1. **Collaborative Problem Solving**:
    - **Prompt**: "Generate a series of math problems on quadratic equations that we can solve as a class. Include guiding questions that I can ask to promote student thinking and engagement."

2. **Discussion Questions**:

- **Prompt**: "Provide discussion questions for a class reading of 'To Kill a Mockingbird' that will help students explore the themes of justice and racial inequality through guided group discussions."

3. **Interactive Activity**:
    - **Prompt**: "Create an interactive activity for a high school history class that involves students working together to map the major battles of World War II, including what each student needs to contribute."

You Do: Independent Practice

1. **Homework Assignment**:
    - **Prompt**: "Design a homework assignment for a physics class covering Newton's Laws of Motion. Include problems that require students to apply each law in real-world scenarios."
2. **Assessment Task**:
    - **Prompt**: "Create an assessment task for an elementary class learning about fractions. Ensure the task includes both multiple-choice and short answer questions to assess understanding comprehensively."
3. **Project Proposal**:

- **Prompt**: "Suggest a project idea for students in a high school art class focusing on the Renaissance period. Outline the project requirements and evaluation criteria."

Creating Worksheets and Slide Show Templates for "I Do, We Do, You Do"

When using the "I Do, We Do, You Do" model, creating supportive materials like worksheets or slide show templates can greatly enhance the learning process. These resources guide students through each phase of the lesson, ensuring that they engage with the content effectively and independently.

Creating Worksheets and Slide Show Templates

Worksheets and **slide shows** serve as structured guides that can be filled out or followed during each phase of the lesson. Here's a step-by-step process on how to create these resources:

1. **Identify Key Points**: Start by identifying the key points and objectives for each phase of the lesson. What do you want students to learn during the "I Do," practice during the "We Do," and demonstrate during the "You Do"?
2. **Structure the Layout**: Organize the worksheet or slide show to mirror the flow of the lesson. Divide it into sections labeled "I Do," "We Do," and "You Do," with clear

instructions or prompts in each section that guide the students' learning.

3. **Incorporate Interactive Elements**: For slideshows, include interactive elements such as polls, questions, or clickable links that make the "We Do" phase collaborative and engaging. For worksheets, include spaces for responses, diagrams for labeling, or boxes for drawing responses.
4. **Provide Guidance and Examples**: In the "I Do" section, provide examples or detailed steps that students can refer to during their independent work. This could be a solved problem, a filled-out diagram, or a list of key terms with definitions.
5. **Design for Differentiation**: Include optional challenge questions or additional resources for students who might need extra help or those who need more advanced materials.

Sample Prompts for Creating Worksheets and Slide Shows Using Chat GPT

1. **Worksheet Creation for a Math Lesson on Algebra**:
    - **Prompt**: "Generate a detailed worksheet for an algebra lesson using the 'I Do, We Do, You Do' model. Include step-by-step problem-solving in the 'I Do' section, collaborative practice problems

for the 'We Do' section, and individual challenge questions for the 'You Do' section."

2. **Slideshow Template for a Science Lesson on Ecosystems**:
   - **Prompt**: "Create a slideshow template for a lesson on ecosystems. Include diagrams for the 'I Do' section, interactive poll questions for the 'We Do' section to discuss ecosystem interactions, and scenario-based questions for the 'You Do' section where students predict outcomes based on changes in the ecosystem."

3. **Worksheet Series for a History Lesson on the American Revolution**:
   - **Prompt**: "Develop a series of worksheets for a lesson on the American Revolution, tailored to the 'I Do, We Do, You Do' model. Incorporate primary source analysis in the 'I Do' section, group discussion points in the 'We Do' section, and a written reflection task in the 'You Do' section."

4. **SlideShow for a Literature Lesson on Shakespeare's 'Macbeth'**:
   - **Prompt**: "Design a slide show for studying 'Macbeth' that includes thematic exploration for the 'I Do', character analysis activities for the 'We Do', and creative writing prompts for the 'You Do'

1. **Single-Question Focused**: These ask a specific question about the day's material, requiring a short, focused answer.

2. **Open-Ended Reflection**: These prompts allow students to comment on what they learned, what they found challenging, or questions they might still have.

3. **True/False or Multiple Choice**: These provide a quick way to assess students' retention of factual information.

4. **Opinion-Based**: These ask students to express their views on a discussion point from the lesson, often used in humanities subjects.

5. **3-2-1 Exit Ticket**: Students list three things they learned, two things they found interesting, and one question they still have.

Using Chat GPT to Create and Personalize Exit Tickets

Utilizing Chat GPT can streamline the creation of personalized and effective exit tickets. Here's how Chat GPT can assist with each type of exit ticket:

1. **Single-Question Focused**
   - **Prompt**: "Generate a concise, single-question exit ticket for a lesson on the Pythagorean theorem

that assesses students' ability to apply the formula."

2. **Open-Ended Reflection**
   - **Prompt**: "Create an open-ended exit ticket prompt that encourages students to reflect on their understanding of the symbolism used in 'The Great Gatsby.'"
3. **True/False or Multiple Choice**
   - **Prompt**: "Develop a five-question multiple-choice exit ticket for a unit on photosynthesis, focusing on the key processes involved."
4. **Opinion-Based**
   - **Prompt**: "Formulate an opinion-based exit ticket question related to the ethical implications of genetic engineering discussed in today's biology class."
5. **3-2-1 Exit Ticket**
   - **Prompt**: "Design a 3-2-1 exit ticket for the end of a world history class covering the Industrial Revolution, tailored to assess comprehension and interest."

These prompts can be modified and reused for different lessons and subjects, making Chat GPT a versatile tool in the educator's toolkit.

CHAPTER 7: ACCOMMODATIONS AND DIFFERENTIATION

Understanding Accommodations and Differentiation

Accommodations and differentiation are essential practices in inclusive education, ensuring that all students have access to the same educational opportunities in a way that considers their individual learning needs. **Accommodations** involve changes in how a student learns material, such as providing extended time for tests or delivering content through auditory means instead of written. **Differentiation**, on the other hand, involves adjusting the instruction, content, or learning environment to cater to the diverse learning styles and abilities of students within a classroom.

How to Identify and Address Diverse Student Needs

Identifying and addressing diverse student needs starts with understanding each student's unique learning profile, which can include their strengths, challenges, preferences, and any disabilities. Teachers can gather this information through formal assessments, observations, student feedback, and discussions with other educators and parents. Once needs are identified, teachers can implement targeted strategies to meet these needs, ensuring that all students are engaged and supported in their learning.

Using Chat GPT to Generate Differentiated Instruction Plans

Chat GPT can assist educators in creating differentiated instruction plans by generating customized content and activities suited to various learning needs. Here's how educators can use Chat GPT:

1. **Individualized Learning Objectives**: Chat GPT can help tailor learning objectives to match different student abilities within a classroom.
2. **Varied Instructional Materials**: Request Chat GPT to provide resources that present information in multiple formats (text, audio, video) to cater to different learning preferences.
3. **Tiered Activities**: Generate activities that are tiered by complexity, allowing students to engage at a level that matches their skills.
4. **Assessment Options**: Use Chat GPT to design various assessment methods that accommodate different learning and expression styles.

Examples of Accommodations for Various Learning Needs

Here are some accommodations that can be implemented, along with corresponding Chat GPT prompts to create materials or strategies for these accommodations:

1. **Extended Time**
    - **Prompt**: Create a guideline for adjusting classroom and testing time limits for students with processing delays.

2. **Visual Aids**
    - **Prompt**: Generate a list of visual aids that can be used to support learning about cellular structures in a biology class.

3. **Simplified Text**
    - **Prompt**: Provide a simplified version of the instructions for a chemistry lab, ensuring clarity and readability for students with reading difficulties.

4. **Alternative Assessment Formats**
    - **Prompt**: Design a project-based assessment alternative for a unit on World War II that allows

students to demonstrate u
models or presentations in

5. **Behavioral Supports**
    - **Prompt**: "Outline a plan for behavioral supports in a cla students with ADHD."

6. **Technology Integration**
    - **Prompt**: List educational te aid students with dyslexia i content, such as text-to-sp audiobooks.

7. **Preferential Seating**
    - **Prompt**: Develop a seating c learning for students with a placing them away from dist the teacher.

8. **Multisensory Learning Opportuni**
    - **Prompt**: Create a multisens teaching fractions that invol auditory components.

Creatir
compo
Ringer
tickets
strateg
ensure
educat
all stu

1.

CHAPTER 8: NTEGRATING ALL ELEMENTS INTO A COHESIVE LESSON PLAN

g a cohesive lesson plan involves integrating various ıents—standards, templates, scope and sequence, Bell , instructional models like "I Do, We Do, You Do", exit and accommodations and differentiation ıes—into a seamless unit of instruction. This integration ; that each part of the lesson contributes to a unified onal objective, enhancing the learning experience for ents.

tart with the Standards: Begin by identifying the tandards that the lesson needs to address. This ensures hat the objectives of the lesson align with educational xpectations and provide a solid foundation for content levelopment.

2. **Incorporate the Template**: Utilize your school's lesson planning template to organize the lesson systematically. This might include sections for objectives, materials, procedures, assessments, and reflections.

3. **Outline the Scope and Sequence**: Place the lesson within the larger context of your scope and sequence. This helps maintain a clear progression of topics and ensures that each lesson builds on previous knowledge and sets the stage for future learning.

4. **Engage with Bell Ringers**: Kick off the lesson with an engaging Bell Ringer to focus students and activate prior knowledge. This should tie into the main topic of the lesson and set a productive tone for the day.

5. **Apply Instructional Models**: Structure the main teaching portion using an effective instructional model, such as "I Do, We Do, You Do", to scaffold learning and gradually transfer responsibility from the teacher to the student.

6. **Include Differentiated Activities**: Plan activities that cater to diverse learning needs, ensuring all students have access to the content in a manner that best suits their learning styles and abilities.

7. **Plan for Assessment with Exit Tickets**: Conclude with an exit ticket that assesses the understanding of the day's material and provides feedback for both students and teacher. This is crucial for adjusting future lessons based on student needs.

8. **Incorporate Feedback and Reflection**: Reflect on the effectiveness of the lesson based on student outcomes and personal observations. Use this reflection to inform future planning.

Tips for Continuous Improvement Using Chat GPT

To continuously improve your lesson plans using Chat GPT, consider these strategies:

1. **Regular Feedback Input**: After each lesson, input any new student feedback and your own observations into Chat GPT. Ask it to suggest adjustments or improvements to the lesson plans based on this feedback.

2. **Update Templates and Standards**: Periodically review and update the templates and standards you input into Chat GPT. Educational standards can evolve, and templates may need adjustments to better serve your teaching style and student needs.

3. **Explore New Ideas**: Regularly use Chat GPT to generate new ideas for Bell Ringers, activities, and assessments. This helps keep your lessons fresh and engaging.

4. **Personalized Learning Adjustments**: Continually refine the differentiation aspects of your lessons by asking Chat GPT for targeted strategies to meet the emerging needs of your students.

5. **Professional Development**: Use Chat GPT to stay informed about the latest educational research and teaching strategies. Ask it to summarize recent studies or articles that could impact your teaching practice.

6. **Collaboration Enhancement**: Share your use of Chat GPT with colleagues to foster a collaborative environment. Exchange ideas on how best to use AI in lesson planning and discuss strategies that have been particularly effective.

4. **Technical Glitches**: Familiarize yourself with basic troubleshooting steps, such as clearing browser cache or restarting the session. Contact technical support for persistent issues.

Maximizing the Effectiveness of Chat GPT in the Classroom

To maximize the benefits of Chat GPT in education, consider the following strategies:

1. **Regular Updates and Training**: Keep the AI model updated with the latest features and ensure regular professional development for teachers to keep up with new tools and methods.
2. **Feedback Loops**: Establish a system where both students and teachers can provide feedback on the AI's performance and the effectiveness of the lesson content generated. Use this feedback to refine how Chat GPT is utilized in lesson planning.
3. **Collaborative Use**: Encourage collaboration between teachers to share effective prompts and successful AI-generated materials. This can lead to the development of best practices within the educational community.
4. **Ethical Use**: Educate teachers and students about the ethical considerations of using AI in education, including

discussions about bias, reliability, and the appropriate use of technology in learning environments.

5. **Experimentation and Creativity**: Encourage teachers to experiment with different types of prompts and explore creative ways to integrate AI-generated content into their teaching. This can lead to innovative approaches and new learning opportunities.

By understanding and leveraging the advanced features of Chat GPT, educators can enhance their lesson planning and deliver more personalized, engaging, and effective education.

CONCLUSION

Recap of Key Points

This ebook has guided you through the dynamic capabilities of ChatGPT in enhancing the art and science of lesson planning. From the foundational understanding of educational standards in Chapter 1 to the sophisticated integration of AI into daily teaching practices discussed throughout the chapters, we've covered how ChatGPT can revolutionize traditional educational paradigms. By leveraging technology to craft engaging Bell Ringers, structure lessons with the "I Do, We Do, You Do" model, design effective exit tickets, and implement accommodations and differentiation, educators are equipped to deliver more personalized and effective teaching experiences.

Encouragement to Innovate and Adapt Lesson Plans Using Chat GPT

As educators, your commitment to continuous improvement is what molds future generations. I encourage you to embrace ChatGPT not just as a tool, but as a partner in education.

Experiment with integrating AI into your lesson plans, allowing it to enhance your creativity and efficiency. Adapt the examples and strategies provided to fit your unique teaching style and students' needs. Innovation in education often requires stepping out of comfort zones, and ChatGPT offers a safe, scalable, and versatile platform for doing just that. Whether it's through developing interactive content, providing instant feedback, or creating diverse educational materials, the potential for growth and improvement is limitless.

Future Trends in AI and Education

Looking forward, the intersection of AI and education will continue to evolve and expand. We anticipate seeing more personalized learning experiences facilitated by AI, where lesson plans and educational content are tailored to the individual learning styles and needs of each student. AI could also play a significant role in predictive analytics, helping educators anticipate learning difficulties and intervene more effectively. Moreover, as AI technology advances, we can expect it to become more intuitive and aligned with pedagogical goals, thereby becoming an even more integral part of educational ecosystems.

The journey of integrating AI like ChatGPT into your teaching practices is just beginning. As you move forward, keep exploring, adapting, and innovating. The future of education is bright and AI is here to enhance it—making it more engaging,

Art Gathering and descriptions - make sure we have the art and the descriptions

Each group should have a task for each day
Using the following lesson plan format:
National Art Standards for High School

{The next part is copied straight from my lesson plan template}

Identify Need: What do you want students to learn and/or be able to do?
Bell Ringer 10 minutes
Learn: What concept or skill will students investigate and how will they do that?
Develop: How will you clarify / model / give feedback to students to facilitate their learning of the concept or skill?
Apply: How will students apply and practice what they have learned?
Evaluate: What will students do to show mastery of the learning objective(s)?
Create a 4 day lesson plan each lesson is 90 minutes
I may have added the standards in a file but you can copy your standards with a description and upload into the prompt also

CHAT GPT RESPONSE:
For this series of lesson plans, I'll structure them around the

task distribution you've outlined, integrating the use of a gallery app for creating a virtual art gallery. The goal is to engage students in a multifaceted project that combines digital art, marketing, event planning, and project management, culminating in a virtual art show.

4 DAYS OF LESSON PLANS.

Lesson 1: Introduction and Project Kickoff

National Art Standards for High School
- Creating: Conceiving and developing new artistic ideas and work.
- Presenting: Interpreting and sharing artistic work.
- Responding: Understanding and evaluating how the arts convey meaning.
- Connecting: Relating artistic ideas and work with personal meaning and external context.

Identify Need
Students will learn about the roles in a virtual art gallery project, including marketing, event planning, gallery setup, and art gathering with descriptions. They will begin to plan and execute their roles in preparation for the virtual art show.

Bell Ringer (10 minutes)

Ask students to write down what they think goes into setting up a virtual art gallery. Have a few share their thoughts.

Learn (20 minutes)
Introduce the gallery app and explain how it can be used to create a virtual art gallery. Divide students into four groups: Marketing, Event Planning, Gallery Set Up, and Art Gathering and Descriptions. Provide an overview of each group's responsibilities.

Develop (20 minutes)
- Marketing: Discuss elements of effective digital marketing and flier design.
- Event Planning: Outline key components of virtual event planning, including timeline, platform navigation, and attendee engagement.
- Gallery Set Up: Demonstrate how to upload and arrange artwork in Spatial.io, focusing on layout and visitor experience.
- Art Gathering and Descriptions: Guide on sourcing digital art (while respecting copyright) and writing compelling art descriptions.

Apply (30 minutes)
Each group starts on their initial tasks:
- Marketing: Brainstorm and sketch flier ideas. Begin creating a social media plan.

- Event Planning: Draft an event timeline and list features of Spatial.io to leverage.
- Gallery Set Up: Practice uploading sample artworks and arranging them in a mock gallery.
- Art Gathering and Descriptions: List needed artworks and start drafting descriptions for each.

Evaluate (10 minutes)
Each group presents their progress and plans for the next steps. Provide feedback and set goals for the next class.

Lesson 2: Deep Dive and Progress

Bell Ringer (10 minutes)
Discuss what makes a virtual event engaging. Ask students to share experiences they've had with memorable online events and why they stood out.

Learn (20 minutes)
Focus on the specifics of each group's tasks, offering deeper insights and examples:
- Marketing: Explore social media strategies and effective online advertising.
- Event Planning: Discuss virtual event challenges and how to address them.

Lesson 4: Final Preparations and Dress Rehearsal

Bell Ringer (10 minutes)
What have you learned about working as a team on a complex project? Share insights on collaboration and division of labor.

Learn (20 minutes)
Review the importance of final touches and the role of rehearsals in ensuring a smooth event.

Develop (20 minutes)
Final feedback session for

REFLECTION ON THE LESSON:
I did have to tweak this a little, which you ALWAYS should to meet the unique makeup of your class. This plan helped me focus on organizing the show with my students. I did create a worksheet that guided the students through the process.

TYPOGRAPHY LESSON

Lesson 2

So in this example, I have already uploaded my template in an early message in this chat. I have also uploaded a PDF labeled 4A, that has the standards and objectives of this lesson

Prompt:
Create a lesson typography where they create their name us 4A pdf to create lesson using the template previously given

CHAT GPT REPLY:
Here's a 90-minute lesson plan focused on typography, specifically designed for students to explore typefaces and create a unique typographic representation of their own names using principles outlined in the "4A Typography Introduction" PDF. This lesson combines theoretical understanding with hands-on application, allowing students to personalize their learning experience.

Lesson Plan: Typography and Personal Expression

- Objective: Students will understand the fundamental aspects of typography and apply these principles to create a unique typographic design of their own names. This activity aims to enhance students' appreciation for typography as a crucial element of graphic design and their ability to use typography to convey personal style and identity.

Bell Ringer - 10 minutes

- Activity: Display several famous logos that utilize distinctive typography (e.g., Coca-Cola, Google, FedEx). Ask students to identify what makes each typeface effective and how it contributes to the brand's identity.

Learn - 70 minutes

1. Concept Introduction (15 minutes)

- Briefly explain the importance of typography in design, covering basic concepts like typeface, kerning, leading, and tracking. Use visual aids from the PDF to illustrate these points.

- Discuss different typefaces (serif, sans-serif, script, etc.) and when it might be appropriate to use each.

2. Skills Practice (20 minutes)

- Activity: Students will explore various typefaces by browsing through a digital or physical typeface library. They should select a few that resonate with their personal style or the image they want to convey with their name.

3. Sketch Application (20 minutes)

- Activity: Using grid paper, students sketch out their names using the typefaces they've researched. Encourage them to experiment with modifying letterforms, such as changing letter spacing, alignment, size, and incorporating elements like shadow or texture.

4. Art Criticism Activity Writing (15 minutes)

- Activity: Students present their sketches to the class. Each student discusses their design choices and receives feedback. They should articulate how their choices reflect their personal identity or style and how the principles of typography guide their design.

Closure - 5 minutes

- Recap the lesson by highlighting the diversity of typography and its impact on communication and design. Ask students to reflect on how typography can be a powerful tool for self-expression.

- Assign a homework task: Students finalize their typographic designs digitally, using graphic design software, and prepare a brief write-up that describes their typographic decisions.

This lesson plan is designed to not only teach students about typography but also to allow them to explore their creativity and how type can be a form of personal expression. It makes the learning process engaging and personally relevant, while also

Made in the USA
Monee, IL
04 November 2026

67385896R00046